Wu Wenjun

A Biography in Pictures

Wu Wenjun
A Biography in Pictures

Written by:
Cai Tianxin
Zhejiang University, China

Illustrated by:
Li Yanan

Translated by:
Tang Bin
Beihang University, China

World Scientific

NEW JERSEY · LONDON · SINGAPORE · BEIJING · SHANGHAI · HONG KONG · TAIPEI · CHENNAI · TOKYO

Published by

World Scientific Publishing Co. Pte. Ltd.

5 Toh Tuck Link, Singapore 596224

USA office: 27 Warren Street, Suite 401-402, Hackensack, NJ 07601

UK office: 57 Shelton Street, Covent Garden, London WC2H 9HE

Library of Congress Cataloging-in-Publication Data
Names: Cai, Tianxin, 1963–   author. | Li, Yanan (Artist) illustrator.
Title: Wu Wenjun : a biography in pictures / written by Cai Tianxin ; illustrated by Li Yanan.
Other titles: Shu xue jia hua zhuan. English
Description: New Jersey : World Scientific, 2023.
Identifiers: LCCN 2023013753 | ISBN 9789811275951 (hardcover) |
    ISBN 9789811275968 (ebook) | ISBN 9789811275975 (ebook other)
Subjects: LCSH: Wu, Wen-tsün. | Mathematicians--China--Biography. |
    Mathematics, Chinese--History--20th century.
Classification: LCC QA29.W82 C35 2023 | DDC 510/.92 [B]--dc23/eng/20230509
LC record available at https://lccn.loc.gov/2023013753

British Library Cataloguing-in-Publication Data
A catalogue record for this book is available from the British Library.

《数学家画传: 吴文俊》
蔡天新 著; 李亚男 绘
Copyright © 2019 by East China Normal University Press Ltd.
All Rights Reserved.

Copyright © 2023 by World Scientific Publishing Co. Pte. Ltd.

For any available supplementary material, please visit
https://www.worldscientific.com/worldscibooks/10.1142/13401#t=suppl

Contributing Editor: Tang Qi, Chong Daoyang
Funded by B&R Book Program

Typeset by Stallion Press
Email: enquiries@stallionpress.com

Preface

The superior man produces his changes as the leopard does when he changes his spots: their beauty becomes more elegant.

— Hexagram Ge (Revolution), *The Book of Changes* (*Yi Jing*)

Mr. Wu Wenjun was a renowned mathematician. When he graduated from college, China was fighting the War of Resistance Against Japanese Aggression. He worked as a middle school teacher in Shanghai for 6 or 7 years and as a substitute teacher at Hangchow University. Afterwards, he continued his study in France and returned to China after obtaining a doctorate degree. Mr. Wu made significant contributions to the fields of topology, ancient Chinese mathematics, and mathematics mechanization. He was elected as a member (academician) of the Academic Divisions of the Chinese Academy of Sciences (CASAD) when he was 38 and was granted the State Supreme Science and Technology Award in 2000 when the award was set up, along with Yuan Longping, the "father of hybrid rice." However, whenever people praise him as a genius, he would respond with a smile that he was not a smart guy and add that mathematics was a subject for the slow-witted.

Outside of mathematics, Mr. Wu loved light reading and was well-read, so to speak. He was also interested in watching movies and stage plays. With a childlike mindset, he once had a snake wrapped around his body when he was 78, sat on the trunk of an elephant when he was 83, and took a taxi to a shopping mall to watch a movie by himself and afterwards had a Starbucks coffee when he was over 90. As he said in his oral autobiography in later years, "What I remember most in my childhood and teenage years were my father's books and the times I buried myself in them."

Mr. Wu was gentle and modest. Even in the times of turbulence, he was never labeled "reactionary," denounced for "maintaining illicit relations with foreign countries," or shut in a "cowshed." That's why someone called him a man with great wisdom appearing slow-witted. He also went away with his nine-year-old only son to Southern China by train for almost a month-long revolutionary experience exchange. That was in 1969. They went to many places, including the West Lake in Hangzhou, where he had worked as a teacher at an early age. They tasted local snacks in different places while visiting around.

Mr. Wu Wenjun knew and understood everything, and he was very much appreciative too. That was what made him a blessed one. At every critical moment in life, he would be advised and helped by his friends or by unexpected people. Therefore, it's no wonder that Mr. Wu enjoyed longevity. Among the 190 outstanding scientists (including 64 social scientists) who were elected as members of CASAD in the 1950s, he was the last to pass away at the age of 99. His departure represented the end of an era.

I met Mr. Wu once about two decades ago, and I'm still impressed by his demeanor and personality. That was the winter of 1996. Mr. Wu was invited to attend an annual mathematics symposium in Taiwan. I was also invited, and Mr. Wu and I were the only two peers from mainland China at the symposium. He was easy-going. We often chatted with each other in our spare time, and we toured the beautiful Riyuetan Pool (the Pool of Sun and Moon) together. After coming back, we also wrote letters to each other, but unfortunately, I have lost all of them.

On the first anniversary of Mr. Wu's loss, I wrote a long commemorative article to *Mathematical Culture*, which was later included in my essay collection of *Travel with Numbers and Roses* (2018) published by China CITIC Press. The day May 12, 2019, marked the 100th birth anniversary of Mr. Wu. Grand commemorative events were held both in Beijing and Shanghai. I also adapted my article into *Wu Wenjun: A Biography in Pictures* and had it published by East China Normal University Press. The illustrator Ms. Li Yanan has put great effort into this book. Her cousin, Li Dan, was a graduate student of mine who unfortunately died of colon cancer during her graduate studies at Zhejiang University. This book is also in memory of her.

I would also like to thank Mr. Wu's family, colleague and Academician Guo Lei, disciple Professor Gao Xiaoshan, Professor Qu Anjing of Northwest University, and Mr. Wu's other peers in mathematics for their comments and valued suggestions on the images of Mr. Wu in different periods and other characters in the book. My thanks also go to academician Tang Tao in Shenzhen, who encouraged me and sent me the book *Follow Your Own Course: An Oral Autobiography by Wu Wenjun* very early in my work. Personally speaking, this is my first published picture book. If readers are interested, more characters can be found in my book, *The Legends of Mathematics*.

Cai Tianxin
At Xixi, Hangzhou in the spring of 2019

Contents

Mathematics is a subject for the slow-witted.

—Wu Wenjun

1

Childhood in Minhouli Lilong, Shanghai

The hometown of the mathematician Mr. Wu is Zhujiajiao Town in Qingpu County, Jiangsu Province (now Qingpu District in Shanghai). Located in the southwest corner of Shanghai, Qingpu is where the city borders the two provinces of Jiangsu and Zhejiang. It used to be under the jurisdiction of Suzhou, boarding Jiashan County of Jiaxing City, Zhejiang Province in the south. The ancestral home of Mr. Wu is Jiaxing, to be exact. His grandfather was a scholar who never secured an official position. When he later gave up that pursuit, he mainly supported his family by working as a private teacher. One year later, Mr. Wu's grandparents and their family fled from a war to Zhujiajiao Town in Qingpu County.

Zhujiajiao Town is remote and small, so it was seldom affected by war, and thus, many monuments here have been preserved until today. Zhujiajiao is a historically and culturally famous town in China and one of the four well-known ancient towns in Shanghai, second only to Fengjing Town in Jinshan District, which is also adjacent to Zhejiang Province. Compared to the family of Wu's father, the family of his mother was quite well-off and were mainly engaged in small handicrafts. On May 12, 1919, Wu Wenjun was born in Zhujiajiao Town.

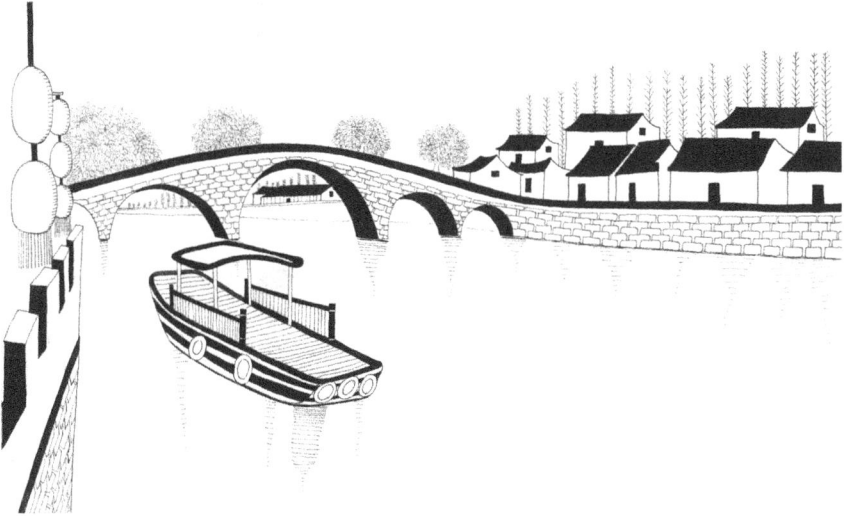

When he was a child, Wu would return to Jiaxing with his family to pay tribute to his ancestors on almost every Qingming Festival. But it seemed that he could not recall the specific locations, except for the memory of "drifting along" to Jiaxing by a hand-crank wooden boat. His father was guileless and studious at a young age. Living in a financially strapped family and as a boy from a poor family background who showed great promise, his father would be funded by his clan or by squires. That was a folk tradition in the regions south of the Yangtze River. It was with the financial support of his grandfather's family that his father was able to enter Nan Yang Public School and finish his preparatory courses.

Nan Yang Public School is the predecessor of Chiao Tung University, and there, Wenjun's father gained a solid foundation in English. After graduation, he worked as a translator in the publishing houses and newspaper offices of Shanghai. At the end of the 19th century, thousands of foreigners lived in Shanghai. At that time, Shanghai was already the most open and most prosperous city in China. The renowned publishing houses of the Commercial Press and the Zhonghua Book Company had already taken shape and had published a large number of classic books. There were many collections in the Wu family, and what impressed Wu the most in his childhood was his father's library.

As far as Wenjun could remember, his home was in Minhouli Lilong (old alleys in Shanghai), Hatong Road (now Tongren Road), Shanghai. Minhouli was an icon of Shikumen (stone gate). Minhouli had a good number of famous artists and writers. Wenjun was the first child in the family. He had a brother and two sisters. His brother Wenjie was smart and adorable; unfortunately, he died as a little boy after some time following a fall from the stairs, though seemed fine at first. As the only boy in the family, Wenjun was showered with even more love by his parents, who wouldn't allow him to play with other kids in the alley as he pleased even when he was a primary school student, just because they were concerned about his safety.

Wenjun's primary school was called Wenwei. This name comes from *The Book of Changes*: "The superior man produces his changes as the leopard does when he changes his spots: their beauty becomes more elegant." It indicates that a superior man should grow as a small leopard does and gradually produce elegant spots—lofty character. Wenjun was fond of reading sophisticated books, such as *The Scholars* (*Rulin Waishi*), and watching films, which made him an open-minded person. He remembered the quote of Liang Qichao, "Heroes only rise in backward countries," which reminded him of the "Prince of Mathematics" Carl Friedrich Gauss (see the picture). When Gauss started his career, mathematics in Germany was still in its infancy. Many mathematicians emerged in Germany later, though another "Gauss" was never seen again.

2

A Goose Egg in a Math Test

Wenjun finished primary school when he was 10, but he repeated a year because his parents thought he was still too young for his grade. In 1930, he went to junior high school at the age of 11. The first year was in Tiehua Middle School, close to his home. The tuition there was costly, yet the quality of education was terrible. The principal would dismiss teachers right after their probation to save expenses. Afterwards, Wu fell ill with a disease resembling typhoid fever. When he recovered, he went to another private school called Minzhi Middle School, located at today's Weihai Road.

In his second year of junior high school, Wu was still studying classical Chinese, particularly *pianwen* (parallel prose), which came into fashion in the Six Dynasties period. The best known *pianwen* was *A Tribute to King Teng's Tower* by Wang Bo, a poet of the early Tang Dynasty. This was a long-standing text in the Chinese Language textbook and was renowned for the line "The autumn water is merged with the boundless sky into one hue." According to records, Qin Jiushao, a great mathematician in the Southern Song Dynasty also liked this literary form. Unexpectedly, during the winter vacation of the second year, the Japanese army started massive bombings of Shanghai. Worrying for the son, Wenjun's parents took him to their hometown, Zhujiajiao, and they hid there for several months.

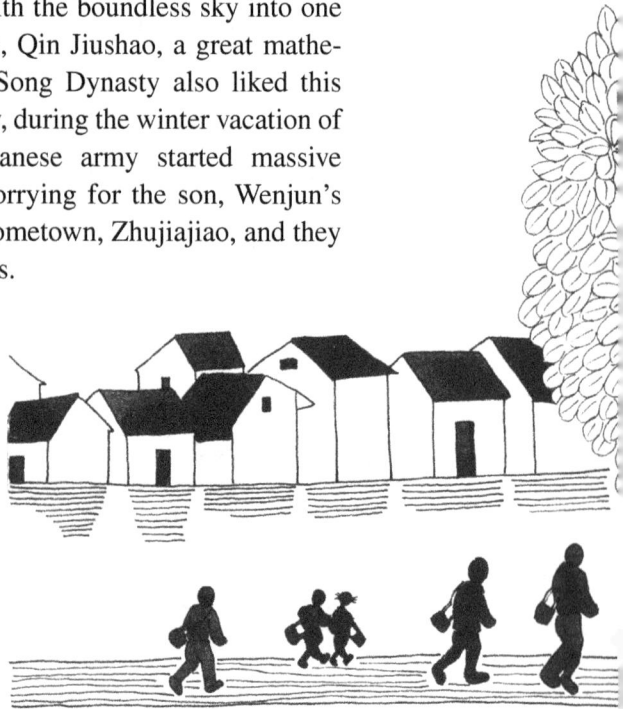

(0 score)

However, classes at the school in the city were not suspended. When Wenjun and his parents returned to the city, he had fallen way behind in his studies. Chinese Language was not a big problem, but mathematics was all Greek to him, so he gave it up and buried his head in novels in class. As a result, he got a goose egg in the final exam. This goose egg was a warning sign to young Wenjun. During the summer vacation, the school began to make up missed classes for the students who had lagged behind because of the bombings. Teachers would ask the students to do problems on the chalkboard one by one and grade their answers right after they finished. This approach really worked, and Wenjun soon mastered the fundamentals and methods of geometry.

In the autumn of 1933, Wenjun went to Zhengshi Middle School, a private senior high school near Xujiahui District. The funder of this school was Du Yuesheng, a gangster tycoon in Shanghai, and the principal was Chen Qun, one of the founding members of the Kuomintang. Comparatively speaking, Wenjun was more relaxed when he was in junior high school, and he really worked hard in the three years of senior high school, especially on math and English. He gained a strong passion for geometry from the experience of making up missed classes during the summer vacation in junior high school and also from his math teacher in senior high school. That teacher's Fujian accent demoralized other students, but it was the other way round for Wenjun.

Wenjun met an excellent English teacher who could play volleyball, from whom he learned to read and write in English. He also read many English novels, including *The Count of Monte Cristo* and *The Three Musketeers* by Alexandre Dumas. As for English listening and speaking skills, they would be acquired at university. It's worth mentioning that Wenjun enjoyed watching stage plays when he was in high school. His favorite actor was Shi Hui, who played the role of Lu Gui in *Thunderstorm*. In this he is a bit like idolaters nowadays. But the difference is that Wenjun watched plays and also read the playscripts, especially those written by playwright Hong Shen.

In addition to mathematics and English, Wenjun's physics score was also high, and he even got full marks in one exam. He was particularly interested in mechanics, even if his hands-on skills were not very strong. However, the physics teacher told the principal that Wenjun's good physics performance was underpinned by his solid mathematics foundation.

So, the principal asked him to sign up for the examination of the Department of Mathematics of Chiao Tung University and promised to award him a scholarship of 100 silver dollars. At that time, the tuition fee of Chiao Tung University was more than 30 silver dollars, which was unaffordable for Wenjun's family, so he had no choice but to agree. One day in an early summer, he received an admission letter from Chiao Tung University.

3

College Life on an Isolated Island

Nan Yang Public School was officially renamed to Chiao Tung University in 1921. In the late Qing Dynasty, Chinese people used to refer to the three northernmost coastal provinces of Liaoning, Hebei, and Shandong as "Beiyang" and Jiangsu and the coastal areas to its south as "Nanyang" (it can be inferred that China's political center of gravity was in the north then). As for the term "Chiao Tung," which is now spelled as "Jiao Tong" in modern Chinese, it was said to also come from *The Book of Changes*, "When the heaven and the earth communicate, all things will be connected." The university anniversary on April 8 also contained the meaning of "Si Tong Ba Da" (extending in all directions). On the other hand, the department in charge of Chiao Tung University in the Republic of China was the Ministry of Communications. At that time, the purview of the Ministry of Communications covered "tanglible communications" and also "abstract and general communications," including trade, diplomacy, postal services, telecommunications, and tourism. Therefore, Chiao Tung University was a comprehensive university.

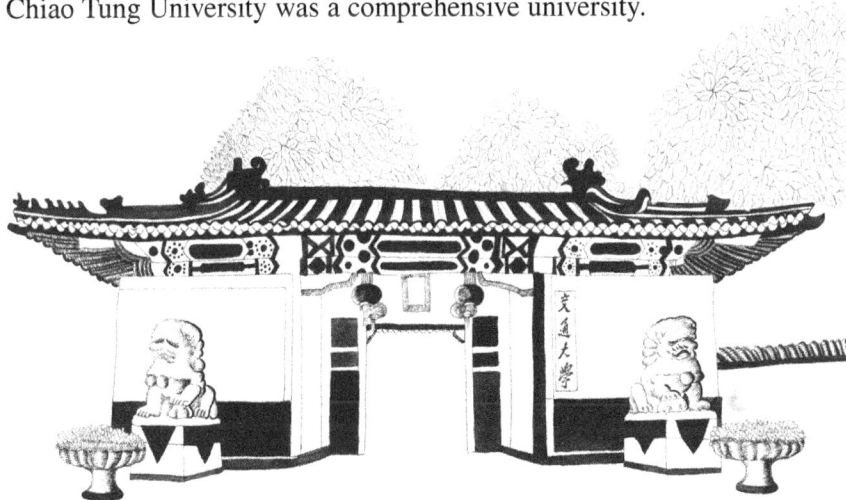

(Chiao Tung University)

Speaking of Nan Yang Public School, China's first Ph.D. in mathematics Hu Mingfu (graduated from Harvard University in 1917; see the left on the next page) graduated from here. In 1910, he as well as Hu Shi, Zhao Yuanren, and other students set off from Shanghai by boat to head for Cornell University in New York State as the second group of students to study in America under the Boxer Indemnity Scholarship Program (the Panama Canal had not yet opened at that time, making their journey long and strenuous). He and Zhao Yuanren studied at the College of Arts and Sciences, while Hu Shi originally studied at the College of Agriculture and then transferred to the College of Arts and Sciences two years later. After returning to China, Hu Mingfu went to Shanghai Utopia University where his elder brother Hu Dunfu (see the middle on the next page) served as the first president and founded and presided over the Department of Mathematics. He was also a part-time professor at his alma mater, Nan Yang Public School. Unfortunately, he was drowned in his hometown Wuxi at the age of 36. This was a loss to the Chinese mathematical world. Otherwise, Wu Wenjun would have met him.

In Mr. Wu's oral autobiography in his later years, he mentioned his calculus teacher Hu Dunfu many times. In 1930, Hu Dunfu stepped down as the president of Utopia University and took up the position of dean of the Department of Mathematics of Chiao Tung University. In 1935, Hu Dunfu joined hands with Xiong Qinglai and Feng Zuxun from Peking University, He Lu from Chongqing University, and Chen Jiangong and Su Buqing from Zhejiang University to establish the Chinese Mathematical Society (CMS). He was elected first chairman of the board of directors. Hu Gangfu (right in the picture above), the third child of Hu's family, was a physicist and one of the first group of returning students from America under the Boxer Indemnity Scholarship Program. He was also a Ph.D. graduate from Harvard University (1918). He was dean of the Faculty of Science of Zhejiang University when it moved westward during the War of Resistance Against Japanese Aggression. When Mr. Zhu Kezhen tried to persuade Hu Gangfu back then, he claimed that if Hu Gangfu didn't take the post of dean, he would not take the office of president of the university either.

Speaking of Wu Wenjun's (see the picture) college life, I read the upper part of the biography of British philosopher Bertrand Russell, *The Spirit of Solitude*, not long ago. It describes Russell's life from his birth to 1921, and he also studied mathematics in college. The Chinese edition has more than eight hundred pages, but the second chapter, "Cambridge," has only 12 pages in total. Coincidentally, the book *Follow Your Own Course: An Oral Autobiography by Wu Wenjun* (interviewed and compiled by Deng Ruohong and Wu Tianjiao, Hunan Education Publishing House, 2015), has more than four hundred pages, and the third chapter, "University Life," also has only 12 pages. A lot of information in this article comes from Mr. Wu's oral accounts. One of the compilers of this autobiography, Wu Tianjiao, is the only son of Mr. Wu. He is currently an engineer at the Academy of Mathematics and Systems Science (AMSS) of the Chinese Academy of Sciences (CAS).

The Faculty of Science of Chiao Tung University was established in 1930 and has three departments: The Departments of Mathematics, Physics and Chemistry. The Department of Mathematics is the smallest, and it only had a total of four students for the first three classes. Wu Wenjun was admitted in 1936. Apart from him, there was another man in the class named Zhao Mengyang from Ningbo, who became his lifelong friend. It was owing to Mr. Zhao's selfless help that some valuable opportunities in life came to Mr. Wu. Because the number of students was small, the three departments would take classes together. In Wu Wenjun's memory, there were four girls in the class. One of them was Lu Zheng from the Department of Chemistry, who was the Gaokao (college entrance examination) top scorer in the Faculty of Science and went to Taiwan after graduation. In the graduation photo of the Faculty of Science in 1940, most students were smiling, whereas Wu Wenjun, wearing glasses and standing at the end of the first row, looked quite bookish.

Wu Wenjun went to classes at Xujiahui, now the Xuhui Campus of Shanghai Chiao Tung University in his freshman year. In his second year of college, the "July 7 Incident" (the Lugouqiao Incident) broke out. After three months of fierce fighting in the Battle of Shanghai, Shanghai fell. Most of the universities in Jiangsu, Zhejiang, and Shanghai moved to Chinese inland areas. The main part of Chiao Tung University also moved to the provisional capital Chongqing, but some still remained in the concessions of Shanghai. Because Wu Wenjun was the only child in the family, his parents let him stay in Shanghai, and the Wu family also moved to the concessions in Shanghai. At that time, there were the Japanese Concession, the French Concession and the British Concession (public concessions) in Shanghai, and the latter two were collectively known as "isolated islands." Chiao Tung University was seated in the French Concession to the south, where Wu Wenjun spent the second semester of his sophomore year as well as his junior and senior years.

Speaking of the French Concession and the British Concession, there was an anecdote saying that the platanus trees (literally "French platanus" in Chinese) on the streets of many cities in China today are not from France, or that there is no such tree with this name in France. It turned out that the platanus was planted in the French and British concessions of Shanghai back then, which is a well-known street tree in the world and enjoys the reputation of "the king of street trees." However, the varieties were slightly different. The platanus in the British Concession had two balls per peduncle, and that in the French Concession had three balls per peduncle. It is hard to say whence it all started, but Shanghailanders refer to them as "British platanus" and "French platanus." The British platanus gradually went into oblivion, probably because the territory and planting area in the British Concession were not as big as those in the French Concession. But the name of French platanus has been used to date and has prevailed throughout southern and northern China.

Because the teaching conditions were tough, life was unsettled, and the teachers were not as good as before on the "isolated island," Wu Wenjun once considered transferring to other departments. But after the Department of Mathematics started separate classes in his junior year, he met Wu Chonglin, an associate professor who taught real variable function theories. Professor Wu guided him through learning patiently and even gave lessons to him at his own home. He also lent him an English book named *Algebraic Geometry* published in India. This course opened the door to modern mathematics for him, and he finally developed a passion for mathematics. Time passed quickly as he pored over classic books of set theory, point-set topology, and algebraic topology, some of which were in German.

In those days, there was a widespread saying in the mathematical community, "Pack your bags for Göttingen." In fact, at the alma mater of Bernhard Riemann and "the Prince of Mathematics," Carl Friedrich Gauss, Felix Klein and David Hilbert had already founded the brand-new Göttingen School. This spurred Wu Wenjun to poured himself into learning German during his studies in Chiao Tung University. In the eyes of every student at Nan Yang Public School or Chiao Tung University, there was a role model named Zhu Gongjin (1902–1962) from Yuyao, Zhejiang Province. Mr. Zhu went to Nan Yang Public School after graduating from primary school and was admitted to Tsinghua University, after which he furthered his study at Göttingen. There, he finished his Ph.D. thesis under the guidance of Richard Courant, one of the favorite students of David Hilbert. After returning to China, Mr. Zhu taught at Chiao Tung University and other schools, and he wrote many articles introducing modern mathematics. Every article was a must-read for Wu Wenjun, who had already learned about the Polish School and the Soviet Union School at that time.

The learning methods of Wu Wenjun were "reading, reflection and taking in." "Reading" means to read the textbooks, "reflection" means to reason the theorems in the textbooks after they are closed, and "taking in" means to master the correlations between all concepts and all theorems. His college thesis paper was titled *The Demonstration of Pascal's Law by Mechanics*. Many years later, he wrote a pamphlet titled *Some Applications of Mechanics in Geometry* (which was published by quite a few publishers) as an extension of his college thesis paper. Hua Luogeng once acclaimed, "This pamphlet is way better than any other ten thesis papers combined." This pamphlet, along with the popular science works of Hua Luogeng, Duan Xuefu, Jiang Boju, and Feng Keqin made up the "Mathematics Book Series," which won the second prize of the National Science and Technology Progress Award (2010). When Wu Wenjun graduated from university, he already had the ambition and confidence to be a mathematician. His teacher, Mr. Wu Chonglin, tried to help him work in the university, but this failed because he had a low status and had never studied abroad, so his words didn't carry much weight.

(Some Applications of Mechanics in Geometry, by Wu Wenjun)

4

Turmoil, Encounter, and Starting Off

In 1940, Wu Wenjun graduated from college at the age of 21. In the following 6 or 7 years, he could have furthered his study or done research, but dark days set in. Wu Wenjun was a teacher in two middle schools in Shanghai. The first one was Yuying Middle School, where he taught algebra with over twenty lessons a week and was also responsible for student management. Every morning, he would run to school to call the roll and see if his students were present for self-study. As for teaching, he had been regretful for not having communicated the basic concept of "two negatives make a positive" until his late years. On December 7, 1941, the Pearl Harbor attack happened. Upon hearing the news, the teachers' office became silent. After a good while, the head of the office sighed, "a nest to be overturned." Shortly after, the Japanese troops occupied the concessions in Shanghai. Yuying Middle School was dissolved, and Wu Wenjun lost his job.

Afterwards, the police stations in Shanghai continued their night patrols and house-to-house checks, but their leader had changed. The Wu family lived in a large room in the French concession. Because Wenjun had two sisters, he slept in the poky attic. One night, the police came to the Wu family on patrol. Staring at the narrow attic set with a bookrack after entering the room, one of the police officers grumbled, "Nothing to check here. This is a family of scholars. They do nothing but reading." Then, they threw a glance at the room and left. The patrolmen that night were Chinese. Though they seemingly submitted to Japanese soldiers, they resisted them deep down. Wu Wenjun detested the Japanese invaders, so he refused to learn Japanese, which he realized later was unwise.

Staying unemployed at home for half a year, Wu Wenjun got a job teaching primary and junior high students at Peizhen School. He still taught the four rules of arithmetic and sometimes assisted in management affairs like calling the roll. In those days, mathematical research was impracticable for him because he was too busy at day, and it was too cramped at home. In addition, he had to sleep early at night because his father needed to get up early for the next day's work. But at school, he and his colleagues got along well, and they talked about anything. One day, other teachers saw him holding a German book and began to keep a distance from him because Germany and Japan were the Axis Powers. Then one day he helped one of them solve a math problem, and they considered him to be on their side again.

In August 1945, Japan surrendered. In the fall semester the same year, Wu Wenjun worked as a substitute teacher in Hangchow University. Hangchow University was one of the 13 missionary universities in the Republic of China period and was dismissed in 1952. This was the first time Wu Wenjun worked in a university. He spent over four months there (today's Zhijiang Campus, Zhejiang University) by the Qiantang River and the Pagoda of Six Harmonies. The dormitory he lived in is called the "White Building" today. It has a direct view of the Qiantang River, whereas the "Red Building" was where foreign professors lived. Among them was Warren Stuart, one of the presidents of Hangchow University. Warren Stuart was the younger brother of John Leighton Stuart, the first president of Yenching University, who was born in Hangzhou and was the last U.S. Ambassador to the Republic of China. John Leighton Stuart also stayed in the "Red Building" in Hangzhou when visiting his brother from Peiping.

In 1916, soon after taking office, Warren Stuart made a decision to set up a bridge over the creeks running between the two mountain slopes on the campus. A century later, the bridge still stands. It's affectionately referred to as "Lovers' Bridge" by the students of Zhejiang University, and many movies have been shot there. At that time, the president of the university was Li Peien, a Hangzhou local. He graduated from the University of Chicago and New York University and taught economics in English. Mathematics was not a prominent subject in Hangchow University, though a contemporary of Wu Wenjun named Zhang Lijing, who studied in Hangchow University, translated many mathematical masterpieces, including the Volume One of *Mathematical Thought from Ancient to Modern Times* (by Morris Kline). Back then, Zhu Shenghao, an alumnus of Hangchow University and the Chinese translator of Shakespeare's works, had already passed away.

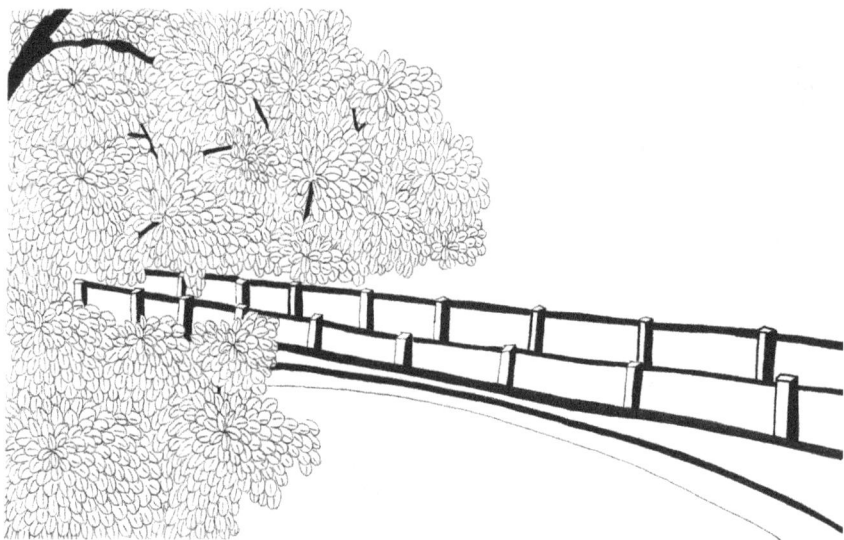

Although Zhejiang University had fostered the formation of the Chen & Su School (a school of differential geometry by renowned mathematicians led by Chen Jiangong and Su Buqing) and had not been relocated from Guizhou to Hangzhou, Zhao Mengyang, Wu Wenjun's schoolfellow at Chiao Tung University, managed to present Wu's thesis paper to the authoritative expert in geometry, Su Buqing. He pulled a few strings with his relatives, probably for the sake of helping Wu secure a job at Zhejiang University. Nevertheless, it was not until many years later that Su Buqing made a comment on Wu's paper, "What an excellent one!" At that moment, Wu Wenjun had already returned from Paris with great accomplishments for a long stay. Zhao Mengyang also managed to introduce Wu Wenjun to the mathematicians Zhu Gongjin and Zhou Weiliang, who had obtained the doctorates of University of Göttingen and Universität Leipzig, respectively, when Wu returned to Shanghai during school holidays. After reviewing an article of Wu Wenjun, Zhou Weiliang remarked, "a sledgehammer to crack a nut." It made Wu Wenjun realize the importance of choosing the right subjects in mathematical research.

In late 1945, the mainstay of Chiao Tung University was still in Chongqing, but it had already set up a makeshift university in Shanghai. It was Mr. Zhao again who generously gave his own job as a teaching assistant in his alma mater away to Wu Wenjun. In spring the next year, the national government announced it would recruit students to study in France by examination, and Zhao Mengyang told Wu Wenjun about it immediately. In the summer of the same year, Zhao Mengyang even took him to visit Shiing-Shen Chern (also Chen Xingshen). In fact, Mr. Zhao didn't know Shiing-Shen Chern, just like he didn't know Zhou Weiliang. What he did was referring Wu to someone who knew them, relying on his ability and passion for making friends. Such meeting resembled the case in the poets' community when a young poet visited a noted one.

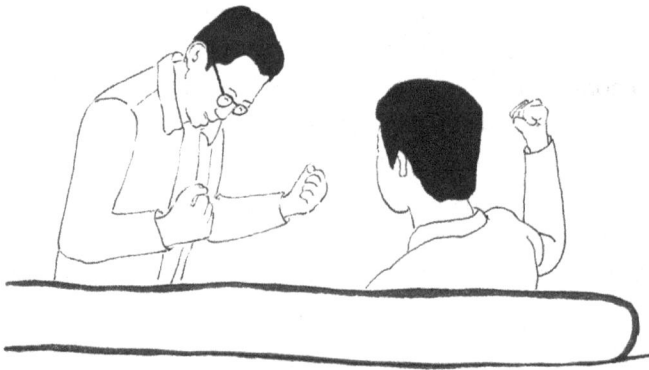

Although Shiing-Shen Chern was only 35, he was far-famed. He made remarkable achievements at Princeton, "The Holy Land of Mathematics," especially his intrinsic proof of the Gauss–Bonnet theorem, and he also worked out the Chern characteristic classes, bringing differential geometry into a new era. After the war was won, he returned to his motherland and planned for building the Institute of Mathematics, Academia Sinica, at the request of his respected teacher Jiang Lifu. Mr. Chern lived in a longtang (alleyway) adjacent to Xujiahui. Wu Wenjun answered his questions when meeting him. Only at the time of parting did Wu Wenjun summon up the courage to ask Mr. Chern whether there were any vacancies in the institute. "I'll keep it in mind," replied Mr. Chern. Not long after, Wu Wenjun got a job at the Institute of Mathematics, located at Yueyang Road close to Fenglinqiao. Since then, he embarked on a royal road to mathematical research.

At the beginning of the building of the Institute of Mathematics, Mr. Chern sent letters to the departments of mathematics in many universities, hoping they could recommend outstanding graduates of within the last three years. In this way, dozens of graduates, primarily from Zhejiang University, came to the Institute of Mathematics. There were also graduates from the National Southwest Associated University, Wuhan University, Sichuan University, Sun Yat-sen University, and Utopia University. Many of them later became the core of Chinese mathematics. In the institute, it was laissez-faire most of the time, except for Mr. Chern's lectures about algebraic topology. Wu Wenjun's office table was right in the library, so he read many books on mathematics. One day, Mr. Chern came to the library and said to Wu Wenjun that he had read enough books and should "pay the debt."

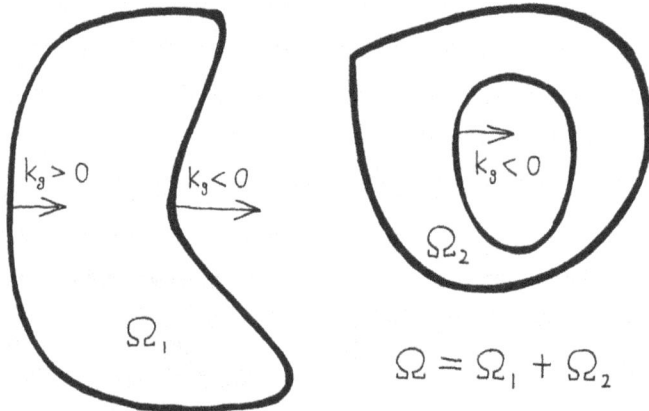

$$\Omega = \Omega_1 + \Omega_2$$

It turned out that by "pay the debt," Mr. Chern meant for Wu to write papers. The first paper Wu Wenjun was pushed into writing was about the imbedding of a symmetric product in Euclidean space, which was published on the *Comptes Rendus* under the recommendation of Mr. Chern. More importantly, Mr. Chern led Wu Wenjun to algebraic topology, which had more space to be opened up than point-set topology, which he was good at. Shiing-Shen Chern was keenly aware of the role of algebraic topology in modern mathematics and its profound impact on other mathematical branches. He believed it would become the mainstream subject of mathematics. In fact, Shiing-Shen Chern took part in inventing differential geometry in the large or global differential geometry world later, the crucial tools of which were fiber bundles and characteristic classes in algebraic topology.

There was a fundamental Whitney product formula in the characteristic classes that had never been rigorously demonstrated, and Mr. Chern hoped that someone could do it when he delivered his lecture. In the spring of 1947, Shiing-Shen Chern made a trip north to give lectures at Tsinghua University, together with two young people, Wu Wenjun and Cao Xihua. Cao Xihua studied at Zhejiang University in Guizhou and obtained his doctoral degree at the University of Michigan in the United States. After returning to China, he taught at Zhejiang University, whereafter he established a new mathematics base at East China Normal University. Making use of his spare time during his stay at Tsinghua University, Wu Wenjun proved Whitney's formula and published it in the leading American journal of mathematics, *Annals of Mathematics*. This was his first major achievement, and he finally began to rise and shine.

5

Glorious Days in France

After Wu Wenjun began to work at the Academia Sinica, he put the spring selection entrance exam to study in France out of his mind. To his surprise, he ranked first in the mathematics group on the list of successful candidates that was published in the second year. Four people passed the exam that year. The other three were Yan Zhida, Tian Fangzeng and Yu Jiarong who graduated from the National Southwest Associated University and Nanjing University. Considering the turbulence in Paris, Shiing-Shen Chern suggested he learn from Henri Cartan in the University of Strasbourg at the France–Germany border, whose father Élie Joseph Cartan was Chern's teacher, and he wrote a recommendation letter for Wu Wenjun.

After finishing two months of training at Nanjing, Wu Wenjun and the other three set off from Shanghai by ship. Taking a completely different route from the other students heading for France, they didn't land at Marseille. Rather, they crossed the Strait of Gibraltar and arrived at Liverpool before they arrived at Calais by way of London and across the English Channel. From there, they took a train to Strasbourg by Paris. Interestingly, when the French official in the education department saw the invitation letter from Cartan in Wu Wenjun's hands, he dispatched all of them to the University of Strasbourg regardless of their majors.

Among them, only Yan Zhida stuck it out to the end. Two years later, he and Wu Wenjun received their doctorate degree together. After returning to China, he taught at Nankai University and was elected as an academician later. Tian Fangzeng and Yu Jiarong later transferred to the University of Paris, and they worked at the Institute of Mathematics, CAS, and Wuhan University respectively after graduation. Although Wu Wenjun was good at both English and German, he didn't master the French language, though he studied in France for four years. He claimed that he, with his rational thinking, was not cut out for the romantic French language, and another possible reason was that he was already 28 years old when he arrived in France.

Wu Wenjun liked the preciseness and rigidity of German, which resembled his personality. Unfortunately, when Wu Wenjun arrived in Strasbourg, Henri Cartan went to the École Normale Super Paris under appointment. Cartan asked his colleague Charles Ehresmann (see the picture), also a student of his father Élie Joseph Cartan, to receive Wu Wenjun. Accidentally, it was a blessing in disguise. Wu Wenjun encountered the right supervisor. Because the subsequent facts showed that his research contents and methods of thinking were different from those of Cartan's, it was not easy to adapt in a short period of time.

While staying in Strasbourg, Wu Wenjun continued his study on characteristic classes in topology. In French tradition, supervisors generally did not set questions for their students in the first place. Instead, they let their students grope their way forward and report to them after achieving some results. After that, the supervisors would provide their incisive advice or even the topic or direction of a doctoral dissertation. At that time, what everyone cared about was whether there were complex structures on manifolds. The necessary condition for the existence of a complex structure was the existence of an approximation of a complex structure.

Proceeding from characteristic classes, Wu Wenjun demonstrated that any manifolds with dimensionalities that were multiples of four had no approximations of complex structures and thus no complex structures. The solution of this problem caused a stir in the academic circles of topology. Heinz Hopf, the top authority on topology at the time, was deeply suspicious, and he even traveled from Zurich, Switzerland to Strasbourg only to see Wu Wenjun. As a result, Hopf was convinced by Wu Wenjun, and he invited Wu Wenjun to pay a visit to Swiss Federal Institute of Technology (ETH Zurich) right away.

In July 1949, Wu Wenjun passed the doctoral dissertation defense in France with *On the Characteristic Classes of Spatial Structure of Spherical Bundles.* He poured himself in studying topology, so he did not tour Strasbourg even though he was about to leave. Many years later, he revisited the place, thinking that he could look at local customs and traditions, yet he was still too occupied to spare the time. Strasbourg is now the seat of the European Parliament. The lights along the city's river lanes at night are fascinating. The University of Strasbourg is also the alma mater of the great German poet, Johann Wolfgang von Goethe.

In the autumn of 1949, Wu Wenjun came to Paris. He studied and worked with Henri Cartan (see the picture above) for two years, which was nearly equivalent to completing a postdoctoral program. He made extraordinary achievements in Paris, known as "Wu's method," making him rise to fame in the academic circles of topology. This was not attributed to the systematic instruction of Mr. Cartan. Though Mr. Cartan, as the founding member and leading figure of the Bourbaki School of Mathematics, surely ranked high in terms of academic achievements, Wu Wenjun didn't embrace the thoughts of the school from beginning to end. Wu Wenjun conjectured that he was not laid off only because he was recommended by Shiing-Shen Chern.

(The Rise and Decline of the Bourbaki School)

Wu Wenjun took part in Mr. Cartan's seminars and also did research on his own. The hotel he rented was in the fifth arrondissement or the Latin Quarter of Paris. It was home to many schools and research institutes, including Sorbonne University, École Normale Super Paris and Institut Henri Poincaré. His room lied semi-underground and was dark by day. A cafe named Sesame Oil in the neighborhood became Wu Wenjun's study. He slept late into the morning, attended academic activities in the afternoon, and pondered mathematical problems in a corner of the cafe at night. He usually would not return to the hotel before the wee hours of the morning.

In the spring of the following year, Wu Wenjun was already honored with great achievements. He and Cartan's other two students, Jean-Pierre Serre (see the second from the left of the picture) and René Thom (see the second from the right of the picture) as well as Heinz Hopf's student Armand Borel (see the first from the right of the picture) were acclaimed as the "four leading topologists" in the field of topology. The sensation caused by the combined work of the four was referred to as a "topological earthquake" in the mathematical community. Among them, Jean-Pierre Serre and René Thom won the highest award in mathematics, the Fields Medal, in 1954 and 1958, respectively.

Jean-Pierre Serre was 27 when he won the prize, and he still holds the record as the youngest winner to date. He was also the first winner of the Abel Prize, established in the 21st century. René Thom referenced many theses of Wu Wenjun in his award-winning work. In Mr. Wu's later years, he still believed that his work in the "topology earthquake" hit the hardest. Meanwhile, some people believed Wu could have won the Fields Medal if he didn't return to China as early as he did. So, what had he accomplished in Paris?

Topology is concerned with the properties of a geometric object that are preserved under continuous deformations. Characteristic classes are fundamental invariants, which are led by Stiefel–Whitney classes, Pontryagin classes and Chern classes. Wu Wenjun named Chern classes and defined Wu classes, making the related calculations easier, which are known as Wu's First Formula. He also revealed the relations among characteristic classes, known as Wu's Second Formula. Cartan highly appraised his work, saying that he was making magical changes.

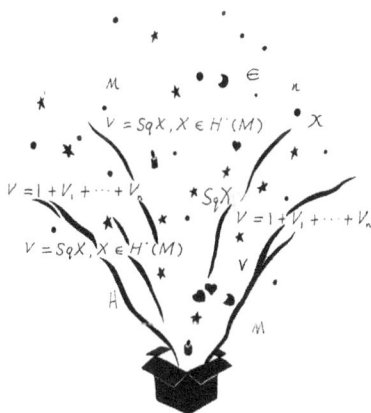

$V = Sq X, X \in H^{\cdot}(M)$

$V = 1 + V_1 + \cdots + V_n$

$Sq X$

$V = 1 + V_1 + \cdots + V_n$

$V = Sq X, X \in H^{\cdot}(M)$

V

6

Return, Honor, and Hover

In the summer of 1951, Wu Wenjun was engaged as a full professor at Princeton University in the U.S. at the age of 32. However, when the letter of appointment arrived in Paris, he was already aboard the ship heading for China. Before that, he was an associate professor at the French National Centre for Scientific Research (Centre national de la recherche scientifique). By that time, China had undergone dramatic changes. The reasons why Wu Wenjun returned to China were various but were primarily related to the traditional culture in which he was raised. That year, he was already 32 years old and unmarried.

The ship departed from the Port of Marseille and crossed the Mediterranean Sea, the Red Sea, the Indian Ocean, and the Strait of Malacca, scheduled to stop at Hong Kong. But when it arrived at Hong Kong, Wu Wenjun, before disembarking, was taken away by the border police and brought to a small steamboat heading directly for Guangzhou. From there, he took a train to his hometown, Shanghai, after years of separation. Later on, Wu Wenjun arrived in Beijing, and his first stop was Peking University. He came on the invitation from Jiang Zehan, founder of Chinese topology and head of the Department of Mathematics, Peking University.

(Peking University)

The year 1952 was a turbulent period for Chinese universities. Wu Wenjun left Peking University to work as a researcher in the Institute of Mathematics, CAS, in a two-story building in Tsinghua Campus. At that time, Hua Luogeng was the director. There were only a dozen people in the institute, including Chen Jiangong and Su Buqing, who were in the far south of the Yangtze River. According to the book *Hua Luogeng* by Wang Yuan, at that time, Guan Zhaozhi was still an associate researcher; Feng Kang was a research assistant; and Lu Qikeng, Wang Guangyin, Ding Xiaqi, Wang Yuan, Gong Sheng, and Hu Hesheng were still interns.

(Tsinghua Garden)

In the spring of the following year, 34-year-old Wu Wenjun went to Shanghai on a business trip where, through the introduction of relatives, he was acquainted with Chen Pihe, a girl who worked in a telecommunications division and knew both English and French. They fell in love at first sight and were married a few days later (half a month later according to some people). A real flash marriage. In the end of the year, his wife was transferred to Beijing to work at the telecommunications division of the Sixth Ministry of Machine-Building, and she then worked in the library of the Institute of Mathematics. She would help Wu Wenjun print foreign theses in her spare time. Later, she gave birth to four children and shouldered the burden of all the chores.

In the winter of 1956, the Institute of Mathematics relocated to Xiyuan Hotel, near Beijing Zoo, and it then moved to Zhongguancun two years later. Before 1957, Chinese scientists and humanists enjoyed a relatively peaceful time, and Wu Wenjun continued his topology research. However, he felt perplexed because academic exchanges were only allowed with the Soviet Union and Eastern European countries where topology research was still backward, whereas that in Western Europe and the United States was leaping forward.

The remarks of Jules Henri Poincaré, French mathematician and founder of topology, occurred to Wu Wenjun's mind, "If we want to foresee the future of mathematics, an appropriate way is to study its history and status quo." Then, Wu Wenjun conducted a thorough review of topology and delivered an academic report in the institute. Invariants were extremely difficult, so people changed 1-1 correspondence to n-1 correspondence to simplify it. Against the trend, Wu Wenjun began to study embedding problems. He introduced the notion of embedding classes and established a theory of embedding.

$$
\begin{array}{ccccccc}
\xrightarrow{\delta_{\bar{p}}} & H^n(img\ p) & \xrightarrow{\delta_p} & H^{n+1}(ker\ p) & \xrightarrow{\delta_{\bar{p}}} & H^{n+2}(img\ p) & \xrightarrow{\delta_p} \\
& n^*\uparrow & & \bar{\pi}^*\uparrow & & \pi^*\uparrow & \\
\xrightarrow{\mu_{n-1}} & H^n(\bar{K},G) & \xrightarrow{\mu_n} & H^{n+1}(\bar{K},\oplus_p^0 C) & \xrightarrow{\mu_{n+1}} & H^{n+2}(\bar{K},G) & \xrightarrow{\mu_{n+2}}
\end{array}
$$

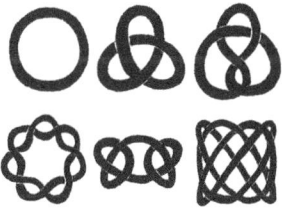

$$
\begin{array}{ccc}
\bar{X} & \xrightarrow{\bar{f}} & \bar{Y} \\
\pi X\downarrow & & \downarrow \pi Y \\
\bar{X} & \xrightarrow{\bar{f}} & \bar{Y}
\end{array}
$$

In early 1957, CAS published its first science awards, the predecessor of the national top three natural science awards of today. The awards were granted to 34 projects, including three first prizes, namely, the "Theory of Functions of Several Complex Variables over Classical Domains" by Hua Luogeng, "Studies on characteristic classes and embedding classes" by Wu Wenjun, and "Engineering Cybernetics" by Qian Xuesen. Mr. Wu's work was recognized in the international topology community. His winning of first prize was unanimous. Significantly, he won the award by eight papers, which is exactly the number of papers eligible for applying for the National Natural Science Awards today.

Hua Luogeng and Qian Xuesen were celebrated at that time, whereas Wu Wenjun was still obscure in the mathematical community in China. Two months after he won the award, he was added as a member of CASAD. During that period, he visited socialist countries with scientist delegations many times, including Romania, Bulgaria, the Soviet Union, the Democratic Republic of Germany (GDR), and Poland. In late 1957, he finished his lectures in the GDR and returned to Paris, from where he had been away for six years, at the invitation of his supervisor, Charles Ehresmann. On this invitation, he gave lectures at the University of Paris. He also visited his alma mater, the University of Strasbourg.

There were government restrictions on Wu Wenjun's visit to Paris banning him from traveling alone, but the French side only invited him, so he had to find another way. As it happened, there was a Chinese student in Strasbourg who then went to Paris, making them a group of two. Mr. Wu met his old friend André Weil in Paris. This mathematics genius was also the nucleus of the Bourbaki School. His younger sister Simone Weil was a world-renowned philosopher.

André Weil (see picture on the right) was also an old friend of Shiing-Shen Chern. He prefaced *Shiing-Shen Chern Selected Papers* and had four Chinese characters "老马识途" (an old horse knows the way) that were handwritten by Shiing-Shen Chern printed on the title page of his own book on the history of number theory. The prominent Langlands Program we know today was put forward in a letter by Langlands to him (1967). André Weil used to ask Wu Wenjun out to an eatery for meals and a chat. At that time, Weil's interest had switched to the history of number theory. Probably, Wu's interest in the history of mathematics later was fostered therefrom.

Wu's visit extended from the scheduled two months to half a year. He also wrote to apply for another two-month extension so that he could finish and publish his monograph in France. Coincidentally, the International Congress of Mathematicians (ICM) was to be held in Edinburgh in August of that year, and the organizing committee invited him to deliver a 45-minute report, making him the second mathematician to be invited after Hua Luogeng since the founding of New China. However, the CAS foreign affairs bureau did not approve but hastened his return for fear that he would be stranded abroad. The Institute of Mathematics, CAS, also sent a telegram on the behalf of Hua Luogeng requesting France to urge his return to China.

During that period, Wu Wenjun was at a loss without knowing how to decide his future. It turned out that he was more attached to his motherland and family. Seventeen years from then on, Wu Wenjun never went abroad. In early 1968, the Romanian Academy of Science sent a letter to him inviting him to attend the academic conference on the topic of Algebraic Topology and Algebraic Geometry as a "guest of our academy" in the autumn that year. The article submitted by the Institute of Mathematics to CAS put it this way, "Based on the consideration of the institute's revolutionary committee, it has been decided that Wu will not attend this conference who should write back to politely decline the invitation by himself."

7

Horse Racing and "Dragon Wu"

In 1956, the central government of China called for "advancing science," so the first natural science award was set up. Soon after, the Anti-Rightist Campaign swept through China. There was also a quota assigned to the Institute of Mathematics. After graduating from Peking University with a postgraduate degree, Professor Shao Pinzong was assigned to the Institute of Mathematics for participating in the number theory panel consisting of Hua Luogeng, Wang Yuan, and Pan Chengdong (Chen Jingrun was transferred to this panel in the autumn of 1957). To fulfill the quota, Shao's superior, considering that he was kind and honest, asked him whether he could help in filling the quota, and after getting his consent, demoted him to Qufu Normal University till his final years.

Thanks to relating well to other people and his outstanding achievements, Wu Wenjun ducked the crisis. Then came the Great Leap Forward. Hua Luogeng had no choice but to set forth the goal to surpass the U.S. in terms of 12 mathematical problems within a decade. This heroic utterance was published on *People's Daily*. However, it was difficult to achieve the surpassing given the fact that research on pure theories including topology that had reached world-class were banned. Hua Luogeng picked the optimum-seeking method. Wu Wenjun was also left with no choice but to switch to operations research.

For some times after exploration, he set game theory as the target of his research, a subject founded by the great Hungarian mathematician John von Neumann (see the picture). What John von Neumann studied was cooperative game, while the Nobel Economics Prize winners, including John F. Nash Jr., studied non-cooperative game. Wu Wenjun was interested in the latter. In one of his popular science articles, he explained that the story of "Tian Ji's Horse Racing" happened during the Warring States Period by game theory. He concluded it as a finite zero-sum game between the two racers.

In 1961, the political trend in China changed. Many "rightists" were exonerated. The central leadership put forward the "fourteen suggestions for scientific research," bringing fundamental research back on the agenda. In the autumn that year, the Chinese Mathematical Society (CMS) held a conference at the Dragon King Temple in the Summer Palace on the development of number theory, topology,

关于摘掉右派分子帽子的决定

摘字 55 号

根据中共中央 (1978) 十一号文件,

(Official government document from 1978)

and the theory of functions. Back then, analytic number theory was in full swing due to the studies of the Goldbach Conjecture by Chen Jingrun, Wang Yuan, and Pan Chengdong, whereas topology studies lagged far behind, including the embedding theory established by Wu Wenjun.

In 1964, the Four Cleanups Movement began, and the staff of the Institute of Mathematics were dispatched down to the countryside by groups. In 1966, the Cultural Revolution broke out. In 1967, Wu Wenjun, flipping through a magazine in the back corner of a reading room, stumbled on an article about integrated circuits at a criticism session of the Institute of Mathematics. At that time, there were no computers, and wire sawing on silicon wafers (chips) was not easy. Wu Wenjun found the theory of embedding class that he established could solve this problem. It gained him a reputation again and also ensured his safety.

Like this, he changed his research directions four or five times within a decade, making it hard to achieve significant results. Fortunately, CAS founded the University of Science and Technology of China (USTC) in Beijing, following the example of the Soviet Union in 1958. Many scientists became professors. As dean of the Department of Mathematics, Hua Luogeng came up with the "dragon-led" teaching method, advocating that a professor should be designated for teaching and leading a class from grade one through grade five. The "three dragons" were Hua Luogeng himself, Guan Zhaozhi, and Wu Wenjun, known as "Dragon Hua," "Dragon Guan," and "Dragon Wu," respectively.

In addition to calculus, Mr. Wu also taught differential and algebraic geometry at the university. According to Li Wenlin, a student of the "Dragon Wu" class and a famous historian of mathematics, Mr. Wu was rigorous in class. He used to sketch out the gist of a lesson on the chalkboard, including the main contents, theorems, and concepts and thoughts it involved, and he then began his deductions. His writing was arranged from the upper left corner to the lower right of the chalkboard in an orderly manner, and then he would wipe it off and start over again. Mr. Hua was dynamic in class. His chalkboard writing was loosely scattered from the right to the left.

One of the best students in the "Dragon Wu" class was Li Banghe from Wenzhou. He followed Mr. Wu's lead and devoted himself to the study of topology. In the field of differential topology, Li Banghe developed the theory of immersions of manifolds in manifolds, extending a foundational theorem in immersion theory that was originally only suitable for the simplest manifolds to fit arbitrary manifolds. Regrettably, Wang Qiming, who recommended Li Banghe to Mr. Wu, died in a car accident in the U.S. in 1989, and the mathematician Qiu Chengtong in the driver's seat was injured. Wang Qiming was a student of Mr. Wu. He once served as the acting deputy director of the Institute of Mathematics, CAS. Wu believed that if it were not for Wang's untimely death, he would have become a leading figure in China's mathematical community.

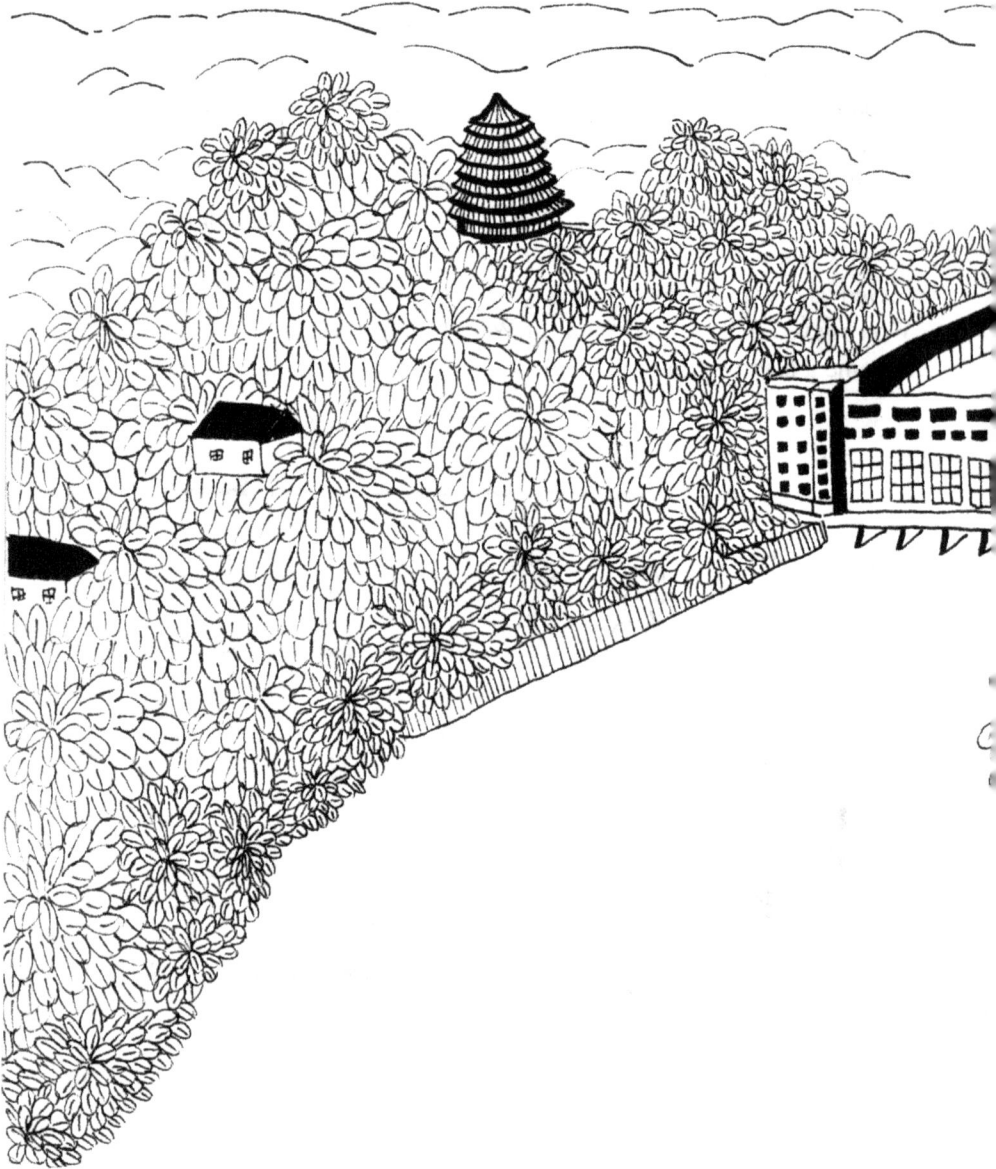

As the Four Cleanups Movement started, teaching was cut off. The first "four cleanups" working crew of the institute went to Jilin, and the second went to Anhui, including Wu Wenjun. In the summer of 1965, they arrived at Sujiabu Town in Lu'an Prefecture, a place sitting next to the Bihe River. The work of Wu Wenjun was to help the production team prepare and fill out reports. He used to do this when he was a middle-school substitute teacher. He ate at a local farmer's home. When he had time, he would go to the market in the town to buy some used novels, which he would read by himself and also lend to others. Once he was stopped by a shout from local people who mistook him for a spy.

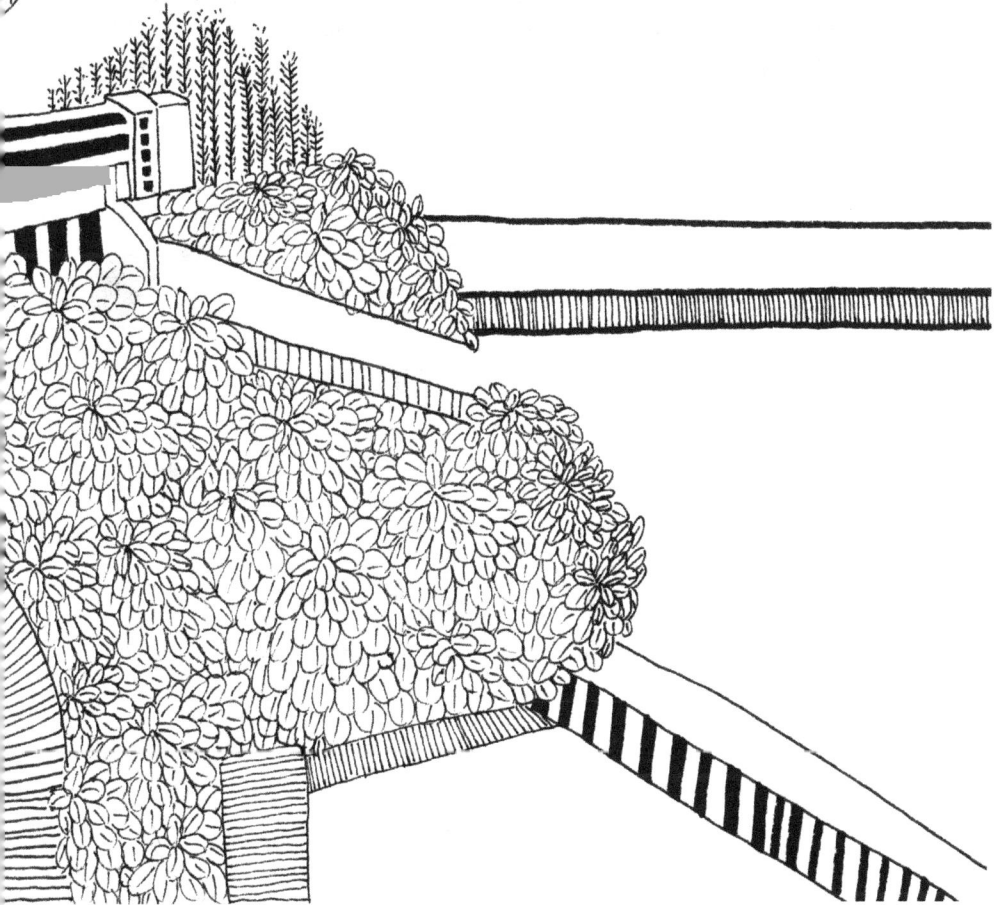

Nearly six months later, the mathematicians were recalled to Beijing because the Cultural Revolution had begun. The "Elimination of the Four Olds (old ideas, old culture, old habits, and old customs)" and seizure of family property followed. The Wu family was searched twice. The first time was mild. He didn't lose much, except for a few light books. The second time was harsh, yet all the seized articles of the Institute of Mathematics were kept on record, so most of them were returned later. Because of fear, many letters were burned by his wife, especially those during his stay in France. Later on, professional books were also abandoned because the room for "academic authority" was shrinking again and again. He went from a four-bedroom and one-living room home to a three-bedroom one and then to a two-bedroom one.

During that time, the home of mathematics master Hua Luogeng, who suffered from serious leg problems then, was also taken over. He was denounced with a nameplate around his neck many times and was also forced to clean the bathrooms of the Institute of Mathematics as punishment, probably because he enjoyed great reputation and liked criticizing others. In contrast, Wu Wenjun had a mild personality and kept on good terms with other people. Although he studied in France for many years and "refused to return," he was not labeled "reactionary" nor denounced for "maintaining illicit relations with foreign countries."

Wu Wenjun was not shut in a "cowshed." He could also go away with his only nine-year-old son to Southern China by a green train for almost month-long revolutionary experience exchanges. They went to many places, including the West Lake in Hangzhou where he had worked as a teacher at an early age. That was in 1969, and the two of them tasted a lot of snacks in the streets and alleys of Hangzhou. Presumably they had also been in Zhiwei Hall or Kuiyuan Hall at that time.

8

Studies on Ancient Chinese Mathematics

The Criticize Lin (Biao), Criticize Confucius Campaign at the later period of the Cultural Revolution kept people away from engaging in academic research; otherwise, they would be denounced as taking the academic path of the bourgeoisie. Since Jiang Qing was leading the tendency towards idolizing the ancients, like wearing the traditional Tang Costume, Guan Zhaozhi, then the deputy director of Institute of Mathematics, came up with the idea of studying ancient Chinese mathematics. Before that, Wu Wenjun didn't have much interest in ancient mathematics and was encouraged to borrow some books about it by Mr. Guan. The first book he read was *The Nine Chapters on the Mathematical Art* (*Jiuzhang Suanshu*). The language in the book was so hard to understand that it was all gobbledygook to him.

(The Nine Chapters on the Mathematical Art)

After that, Wu Wenjun read the books of historians of mathematics Li Yan and Qian Baocong, especially the *History of Chinese Mathematics* by Mr. Qian, the founding dean of the Department of Mathematics of Zhejiang University. Once, Wu Wenjun found the *Jade Mirror of the Four Unknowns* by Zhu Shijie, a mathematician in the Yuan Dynasty, in a second-hand bookstore. There were also two historians of mathematics Li Di from Inner Mongolia Normal University and Li Jimin from Northwest University who helped him considerably, especially Li Jimin's understanding and interpretation of *The Nine Chapters on the Mathematical Art,* which enabled him to truly grasp the charm of ancient Chinese mathematics.

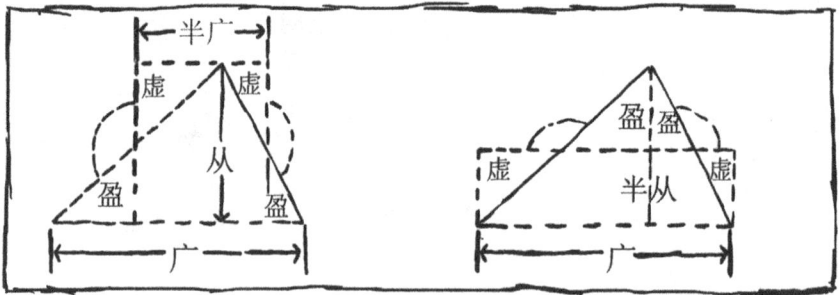

(A geometric problem in *The Nine Chapters on the Mathematical Art*)

The ancestral home of Li Jimin (see picture on the right) was in Xinjin, Sichuan Province. His father once served as a company commander of the Sichuan Army. He was born in Jiujiang, Jiangxi Province. He was sturdy in his youth and liked swimming. He even swam across the Yangtze River once. During his middle school years in Chengdu, he was the chairman of the student union. When he graduated, he failed to enter an elite college due to his family background. In 1958, he entered the Department of Mathematics of Northwest University. During his undergraduate study, he completed a research paper on the theory of functions under the guidance of a professor. As recommended by Hua Luogeng, he became the first person to publish a mathematical paper in the *Science China* journal in Northwest China.

However, Li Jimin failed to stay and work in Northwest University when graduating from the school due to his family background and was assigned to Xi'an Evening University.

During the Cultural Revolution, the evening school was dismissed. Li Jimin was dispatched to Mianxian County, Hanzhong, to work and settle in a production team. In 1972, he appealed to higher authority for help with his published paper in arms but was transferred back to Xi'an Normal School. In 1979, under the care of Vice Premier Fang Yi, he finally returned to his alma mater, Northwest University. Under extremely difficult conditions, he began to study the history of mathematics and made remarkable achievements.

(Northwest University)

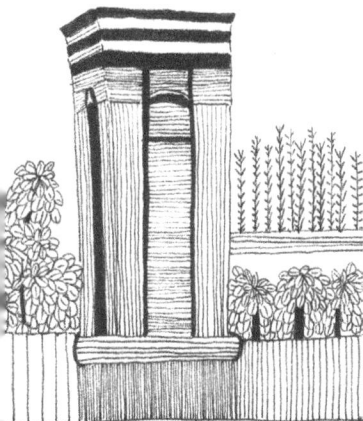

At the turn of the spring and summer of 1977, Wu Wenjun traveled to Xi'an on a one-month northwestern trip. He spent many days with Li Jimin there. After returning to Beijing, he hand-painted a travel map. In Mr. Wu's opinion, Li Jimin was one of the strongest promoters of history of mathematics studies in China after Li Yan, Qian Baocong, and Yan Dunjie. With his support and recommendation, Northwest University set up its first doctoral program on the history of mathematics, but unfortunately, Li Jimin died at a young age. In 2002, Li Jimin's student Qu Anjing delivered a 45-minute report at the International Congress of Mathematicians (ICM) in Beijing on invitation.

Wu Wenjun's first breakthrough was re-demonstrating the solar altitude formula proposed in *Graphic Illustration of Solar Altitude* (*Rigao Tushuo*) by Zhao Shuang, a commentary on *Arithmetical Classic of the Gnomon and the Circular Paths of Heaven* (*Zhoubi Suanjing*). Zhao Shuang was one of the greatest mathematicians in the Three Kingdoms Period (he and Liu Hui from the Wei State pioneered the demonstration of the Pythagorean theorem). His approach was to calculate the sun's altitude by measuring the shadows of two graduated poles of known heights and distance between them erected on a level ground in the city of Luoyang. In 1975, Wu Wenjun reproduced the demonstration completed over two thousand years ago and published it in *Acta Mathematica Sinica* using the pen name of "Gu Jinyong."

Later on, he reproduced the demonstration of calculating the "elevation of sea island" illustrated in *The Sea Island Mathematical Manual* (*Haidao Suanjing*) by Liu Hui. Following *The Nine Chapters on the Mathematical Art*, another zenith of Chinese ancient mathematics arrived when *Mathematical Treatise in Nine Chapters* (*Shushu Jiuzhang*) by Qin Jiushao came out. Qin's method of finding one solution (the Chinese Remainder Theorem) and root-extraction by iterated multiplication (Qin Jiushao's algorithm) were highlights of ancient Chinese mathematics. As Wu Wenjun saw it, Qin's method lied in its constructivity and potential for mechanization. He could solve algebraic equations of high degree with Qin's algorithm by using a small calculator. Furthermore, Qin's method of finding one solution was quite effective and surpassed that of Westerners yet required simpler conditions.

In Wu Wenjun's opinion, algebra is the most extraordinary feat in the history of ancient Chinese mathematics, and *The Nine Chapters on the Mathematical Art* is an encyclopedia of algorithms that embodies the world's earliest geometry and oldest equation sets and matrices. The method of elimination for solving equations in it was much earlier than that of Carl Friedrich Gauss, and it set out the concepts of positive and negative numbers and even real number theory. He also believed that theorem proving was not particularly important for our ancestors; they attached great importance to solving practical problems. They believed there must be certain relations between different data and illustrated these relations through equations.

It can be inferred that the ancient Chinese mathematics Wu Wenjun had been referring to was the period before the 17th century. Since the discovery of calculus, Chinese mathematics noticeably lagged behind the West in terms of calculation. The inspiration Mr. Wu drew from his research of ancient Chinese mathematics had advanced his career in mathematics mechanization. This was also the reason why he firmly believed his decision to return to China was undoubtedly right. In the words of Mr. Fu Ying (see the picture), a chemist who once served as vice president of Peking University, "The history of a branch of science is its gem, because science is a path to knowledge, but history, to wisdom."

Wu Wenjun drew a "Diagram of Mathematical Development" summarizing the viewpoints of Qian Baocong. He held that Western Mathematics has two roots: one was in China and reached the West via India; the other was in Greece and reached the West by way of Arabia. He believed that the greatest mathematic invention of our ancestors is positional notation (or place-value notation or positional numeral system). We had a formal decimal system as early as the Shang Dynasty, the same period whose oracle bone inscriptions served as the basis for Chinese characters. In 1986, Wu Wenjun was invited to give a special report on Recent Studies in the History of Chinese Mathematics at the International Congress of Mathematicians held in Berkeley, California, the U.S.

It's worth mentioning that around the time of the U.S. president Richard Nixon's first trip to China in the latter period of the Cultural Revolution, Chinese scientists such as Yang Zhenning and Shiing-Shen Chern began to return to China. Along with them were also some top U.S. scientists who came to China to give lectures. Against this backdrop, the Institute of Mathematics and the Institute of Systems Science of CAS and Peking University jointly started a seminar on differential geometry. Wu Wenjun resumed his study on topology and made a remarkable comeback.

Wu Wenjun introduced a new fundamental invariant, the I*-measure as well as the new concept of computability, thus enriching the research objectives and skills of topology. This result was later edited by Mr. Wu himself and published by the well-known German publisher Springer Verlag in 1987. It was also included in the long-standing authoritative yellow-cover book *Lecture Notes in Mathematics* (1264).

9

New Proof of Geometry Theorems

Everyone has something to be proud of. When Mr. Wu looked back on his life in later years, he put his study on topology in the third place, above that, ancient Chinese mathematics. Perhaps someone might say, this is an application of the game theory because his research on topology has long been recognized by the world, and topology per se is a symbol of elegance and sophistication in mathematics. Anyhow, it could be inferred that what Mr. Wu was most proud of was his research on mathematics mechanization in his later years and what he had achieved in this area was undoubtedly at the top of the list in his mind.

Now, let's talk about Guan Zhaozhi, Wu Wenjun's most trusted colleague and leader. Mr. Guan is a native of Tianjin. He graduated from Yenching University and continued his studies in France with Mr. Wu. He studied mathematics in Paris, though he was originally heading for Switzerland to study philosophy. Being an underground Party member, he returned to China with his Ph.D. studies unfinished after the founding of New China. It was him who inspired Wu Wenjun to publish Chen Jingrun's research results on Goldbach conjecture. In 1980, Guan Zhaozhi was elected as a member of CASAD. He served as the leader of the Institute of Mathematics for a long time and was also distinguished in the academic field.

(Institute of System Science)

Guan Zhaozhi founded the Cybernetics Research Office under the Institute of Mathematics, the predecessor of the Institute of Systems Science (ISS) that was founded in 1979. The announcement of establishing ISS was made by Qian Sanqiang (vice president of CAS), when Director of the Institute of Mathematics Hua Luogeng and Secretary of the Institute of Mathematics Wu Xinmou were abroad. Guan Zhaozhi and Wu Wenjun were appointed director and deputy director of ISS. After Mr. Hua returned to China, he asked someone to meet Wu Wenjun in the hope that Mr. Wu could return to the Institute of Mathematics. Some experts even held that mechanical proving was deviation from the norm. But Guan Zhaozhi was in support of him, declaring that "Let Wu Wenjun do whatever he wants."

Born in Jiangyin, Jiangsu Province, Wu Xinmou was the founder of the partial differential equation studies in China. He and Hua Luogeng were the same age. He graduated from National Central University and furthered his study in France at a time earlier than Guan Zhaozhi and Wu Wenjun. He married a French woman and brought her to China. They had many children, among which were a pair of twin brothers Wenbei and Wenzhong. Naturally, these two brothers were easy to be noticed among the children of scientists of CAS at that time.

In the 1980s, Guan Zhaozhi, Hua Luogeng, and Wu Xinmou passed away one after another. In 1998, the Academy of Mathematics and Systems Science (AMSS) of CAS was founded with the integration of the Institute of Mathematics, the Institute of Systems Science, the Institute of Applied Mathematics, and the Institute of Computational Mathematics. Academician Yang Le, a function theory expert, became the first president of AMSS, followed by academician Guo Lei, a systems scientist, and algebraist Xi Nanhua.

(Academy of Mathematics and Systems Science)

Dating back to 1971, the "three orientations" campaign towards factories, rural areas, and schools was put forward. With experience as a middle school teacher, Wu Wenjun volunteered to go to factories during this time. When he came to Beijing No. 1 Radio Factory, he saw a computer for the first time in his life. After entering a few numbers and pressing a few keys, the solution and curve of a differential equation showed up all at once. Wu Wenjun was astounded. In fact, in 1948, shortly after the invention of the computer, Polish-born Alfred Tarski already theoretically proved that any propositions could be decided in a mechanical way within the scope of elementary algebra and geometry.

In 1959, Chinese American mathematician Wang Hao, who graduated from National Southwest Associated University, successfully proved hundreds of logical propositions in *Principia Mathematica* by Alfred North Whitehead and Bertrand Russell on an IBM computer. It only took him three minutes. He verified the feasibility of using computers to prove theorems for the first time and introduced the notion of "mathematics mechanization." However, the mechanical proving of geometry theorems was much more complicated than logical propositions. Despite their advanced computers, scientists in the U.S. failed in the mechanical proving of geometry theorems using Alfred Tarski's method.

Inspired by the thought of René Descartes, Wu Wenjun transformed geometry problems into algebra problems by introducing coordinates before mechanizing them. Generally speaking, geometry theorems are deduced from hypotheses. Whether it's a hypothesis or a conclusion, a set of formulas can be deduced after coordinates are introduced, which can be respectively referred to as hypothetical equations and conclusive equations. The kernel of mechanical proving is to seek the solutions of hypothetical equations that also solve the conclusive equations, while giving logical geometric interpretations.

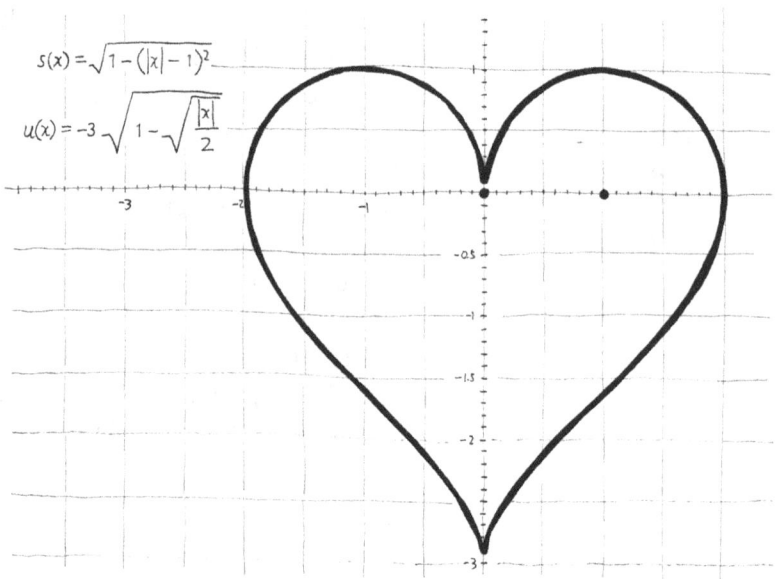

$$s(x) = \sqrt{1 - (|x| - 1)^2}$$

$$u(x) = -3\sqrt{1 - \sqrt{\frac{|x|}{2}}}$$

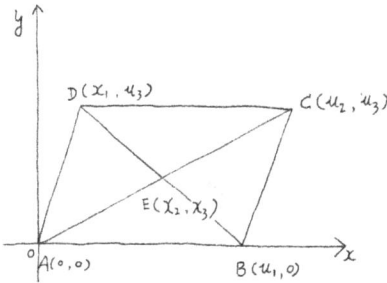

$$\begin{cases} f_1^{\bullet}(u_1,u_2,\cdots,u_n,x_1^{\bullet})=0, \\ f_2^{\bullet}(u_1,u_2,\cdots,u_n,x_1^{\bullet},x_2^{\bullet})=0, \\ \cdots\cdots \\ f_m^{\bullet}(u_1,u_2,\cdots,u_n,x_1^{\bullet}x_2^{\bullet},\cdots,x_m^{\bullet})=0. \end{cases}$$

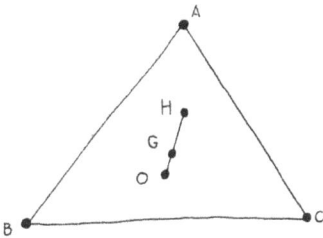

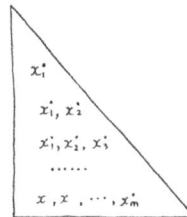

Wu's method for the mechanical proving of geometry theorems is an approach to constructive algebraic geometry. Classical algebraic geometry studies the zeros of multivariate polynomials, whereas modern algebraic geometry uses abstract algebraic techniques to solve geometric problems related to sets of zeros. When he was studying at Chiao Tung University, Wu Wenjun read *Algebraic Geometry*. He opened a course on this at USTC, teaching and researching while studying. In 1977, Wu Wenjun verified his method for mechanically proving elementary theorems of geometry by hand computation. The following year, he applied this method to mechanical proving in differential geometry.

To prove more geometry theorems surely entails a computer. The better the computer performance, the more profound the geometry theorems that can be proved. The first computer Mr. Wu used was an Italian-produced, hand-cranked mechanical calculator. Then, he had a domestically produced computer and a pocket calculator a foreigner who visited the institute gave him as a gift. Later, he got a U.S.-made computer for US$25,000. This was owed to academician Wang Dezhao, a physicist of underwater acoustics who studied in France. He revealed to Mr. Wu the schedule of Li Chang, vice president of CAS.

After having a computer, Mr. Wu had to learn how to program. Mr. Wu began learning programming at almost the age of 60. He managed to master various computer languages from Basic to Algol and Fortran. When computer rooms appeared in the 1980s, the time of computer use needed to be allocated. It was found that the time Mr. Wu spent on computers in those years was far ahead of others in the institute. As academician Wang Xuan, the inventor of laser photo-typesetting system of Chinese characters, once recalled, he saw Mr. Wu still working on a computer in the computer room on a New Year's Eve while he was taking a walk outside the teaching building.

Following the success of Wu's method, a number of mathematical theorems were proved. The method even benefited Zhou Xianqing, a student from the Department of Computer Science who attended the classes Mr. Wu gave to Ph.D. students at USTC as a guest student. Zhou Xianqing later went to the University of Texas, where the dean of his department was working on mechanical theorem proving but without much success. Zhou introduced Wu's method to them, and the method worked so well that it was soon widely known throughout America. In 1990, the Mathematics Mechanization Research Center was established with support from the State Scientific and Technological Commission of the People's Republic of China, and the research fund approved was 1 million yuan. Wu Wenjun was appointed the director of the Research Center. Later, its "Mathematical Mechanization Automatic Inference Platform" became one of the first projects included in the National Basic Research Program of China (973 Program).

10

Later Years: Glory and Peace

In 2001, Wu Wenjun and the "father of hybrid rice," Yuan Longping, won the first State Supreme Science and Technology Award. This was credited to his achievements and also his age. That year, Mr. Wu was 82 years old and healthy. Mr. Wu "cottoned up" to Mr. Yuan, saying, "Agriculture and mathematics are always closely related, and mathematics originated from agriculture." This is indeed general knowledge of mathematics in view of the fact that mathematics appeared during shepherds' calculating the number of their sheep. Prior to this, Mr. Wu was granted the Third World Academy of Sciences (TWAS) Award, the Tan Kah Kee Science Award, the first Qiu Shi Science and Technologies Foundation's Outstanding Scientist Award, and the Herbrand Award for Distinguished Contributions to Automated Reasoning.

In 2006, he received US$1 million for the Shaw Prize in Mathematical Sciences in Hong Kong, which was one of the awards he was most proud of. This science award, established by Hong Kong film & TV producer Shao Yifu (Sir Run Run Shaw), is acclaimed as the "oriental Nobel Award." Two of its three categories of awards, namely mathematics and astronomy, are not included in the Nobel Prize. This was said to be an idea of Mr. Yang Zhenning (Chen-Ning Yang), the designer of this award. For the mathematics category, Wu Wenjun served as the chairman of the Shaw Prize Council for the first and second prizes, awarded in 2004 and 2005, respectively, at the invitation of Yang Zhenning.

The first Shaw Prize in Mathematical Sciences went to Shiing-Shen Chern, and the second went to British mathematician Andrew Wiles, who proved Fermat's last theorem. The chairman of the council of the third Shaw Prize in Mathematical Sciences was Lebanon-born British mathematician Michael Atiyah, with four mathematician judges from China, the United States, Japan, and Russia. The winners were Wu Wenjun and American mathematician David Bryant Mumford (see the first person on the left of the picture). They all shifted their focus of study from the traditional mathematics of algebraic geometry and topology to the new computer-related fields, which "signaled the future development direction of mathematics and also provided a new model for other mathematicians."

Since 1979, Wu Wenjun embarked on a journey of giving lectures in the United States, Canada, East and West Germany, Italy, Switzerland, South Korea, Singapore, Australia, and other countries. He also went to Paris and Strasbourg multiple times to reunite with his former supervisor, classmates, and friends. In the winter of 1996, Mr. Wu was invited to attend an annual mathematics conference in Taiwan. I was also invited, and Mr. Wu and I were the only two peers from mainland China at the symposium. Mr. Wu was easy-going. We often chatted with each other in our spare time, and we toured the beautiful Riyuetan Pool (the Pool of Sun and Moon) together.

The following year, I applied for the Fok Ying-Tong Education Foundation for Young Teachers with my work titled "Mathematics and Art." When I made a call to Mr. Wu asking whether he would write a recommendation letter for me, he agreed, to my surprise. I remember that Mr. Wu's letter was written in blue ink, conveying the ideas that mathematics and art are related in many ways, which, however, people usually turn a blind eye to. Now that Mr. Cai (the author) has blazed a trail, it's worth supporting, and so on. Unfortunately, my application was declined. Since then, I have never seen him again. Twenty years later, the Ministry of Education set up a special fund for popular science publications. Finally, my application got accepted, which could probably bring some comfort to Mr. Wu.

(Legends of Mathematics: Those Great Mathematicians, by Cai Tianxin)

In 1984, Wu Wenjun was elected chairman of the Chinese Mathematical Society (CMS). Due to his efforts, the Chinese Mathematical Society (CMS) became a member of the International Mathematical Union (IMU). Meanwhile, he didn't turn a cold shoulder to the Mathematical Society in Taipei. Both of them held a place as parts of the integral whole: China. In the wake of the 50th anniversary of CMS, Mr. Wu invited his supervisor Henri Cartan to visit China and kept him company on the tour of the West Lake. Mr. Wu attributed his thoughts on mechanical proving to his supervisor. In those years, he could barely keep up with his supervisor in class. As the class went on to a certain point, it became "mechanized" step by step in a cut-and-dried manner.

Before stepping down, Wu Wenjun worked to have established a non-consecutive-term system for the chairman of CMS, and this tradition has continued to this day. In 2002, the International Congress of Mathematicians (ICM) was held in Beijing, and Mr. Wu served as the chairman of the congress. He quoted Napoleon Bonaparte at the opening ceremony: "The progress and improvement of mathematics is linked to the prosperity of the state." He also delivered a speech titled "The Real Number System in Ancient Chinese Mathematics." John F. Nash Jr., the subject of the movie "A Beautiful Mind," also attended the conference and made a speech. In 2001, he set up the Mathematics, Astronomy, and Silk Road Fund with 1 million yuan for the prize money.

As head of the academic leadership group of the Tianyuan Fund for Mathematics, Mr. Wu placed great importance on applications. He was a strong advocate for giving priority to funding research on financial mathematics, a leading figure of which was Professor Peng Shige (see the picture), who enjoys great renown today. With regards to the application of mathematics mechanization, his first choice was numerically controlled machine tools. From parallel to serial ones, they were embedded with core algorithms to improve efficiency. This was done using computer-aided design, which has since become a widely applied technique at home and abroad. Some of Mr. Wu's students have done quite well in this area, such as Gao Xiaoshan. Mr. Wu has got successors, so to speak.

In the eyes of Professor Gao Xiaoshan, mathematics mechanization generally solves mathematical problems with computers as an aid. Mr. Wu is working in algebra mechanization. Although there are also topology mechanization, algebraic geometry mechanization, and others, these are exceptionally challenging. Among them, the most successful one in the world now is the mechanization of number theory, also known as computational number theory, which is directly applied to cryptography. Based on my research experience, computers are to number theorists as telescopes are to astronomers.

Mr. Wu had a childlike mindset. In 1979, he, at the age of 60 in the U.S., was thinking of taking a Greyhound bus to travel across the American continent. In 1997, he, at the age of 78, had a snake wrapped around his body in Australia. In 2002, he, at the age of 83, sat on the trunk of an elephant in Thailand. Meanwhile, Mr. Wu liked quiet places and was interested in light reading and movies. In his 90s, he even escaped the notice of his family, took a taxi to a shopping mall to watch a movie by himself, and had a Starbucks coffee afterwards.

勇攀高峰
再创辉煌

敬贺上海交通大学
110周年校庆

吴文俊

二〇〇六.四.二

Strive for high peaks. More glories to
come. Congratulations on the 110th anniversary of the founding of Shanghai Jiao
Tong University.

Wu Wenjun
April 2, 2006

Once, Mr. Wu stood in the way of a young colleague's car in the institute asking for a lift to the cinema after he failed to hail a taxi. "Mr. Wu is a distinguished person in terms of both manner and scholarship. He is a role model of Chinese mathematicians," remarked Xu Zhongqin of the National Natural Science Foundation of China (NSFC), who worked with Mr. Wu for many years. Professor Cheng Minde of Peking University once praised Mr. Wu, "He is a man of great wisdom appearing slow-witted." Professor Hu Guoding of Nankai University agreed, "He is a man of great wisdom. Although he keeps saying that he does not understand and he has no clue, it is actually the other way around."

Also, Mr. Wu was always grateful. That was what made him a blessed one. At every critical moment in life, he would be advised and helped by his friends or by good-willing strangers who happened to cross his path. In such a complicated and turbulent period, he passed a long life safely. However, whenever people praised him as a genius, he would respond with a smile that he was not a smart guy and add that mathematics was a subject for the slow-witted.